It Does Matter!:

Different States of Matter

(For Kiddie Learners)

SPEEDY
PUBLISHING

Speedy Publishing LLC
40 E. Main St. #1156
Newark, DE 19711
www.speedypublishing.com

MATTER includes everything around us—our couch, our bed, our computer, our food and our drinks our family and our dog or cat the grass and the trees the Sun and the Moon all the planets everything in our Universe.

Yes! Everything!

Matter is anything which occupies space and has mass.

Three common states of Matter are solid, liquid, and gas.

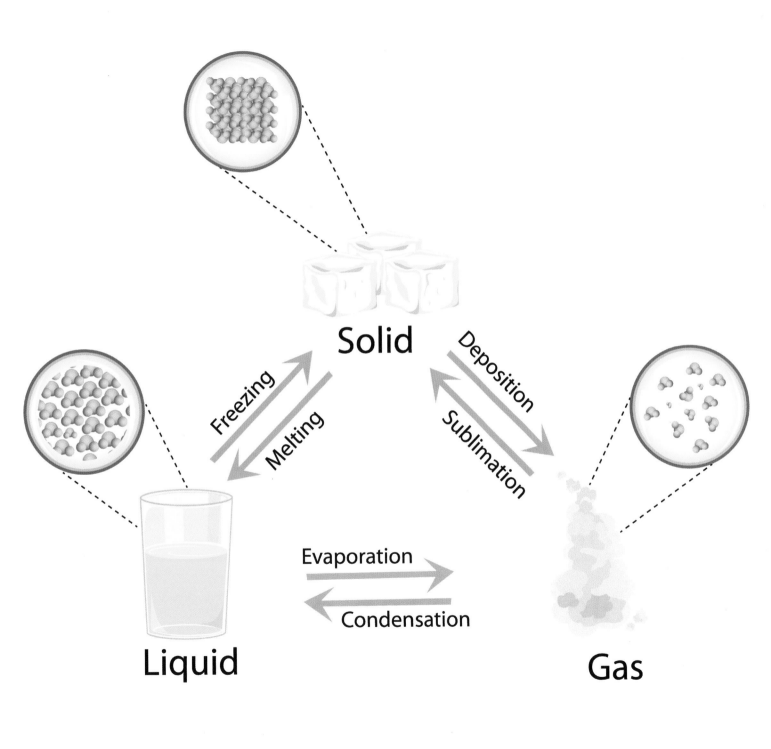

The properties of these states can be explained in terms of the kinetic theory of matter.

Matter's made up of small particles. The particles are in constant motion.

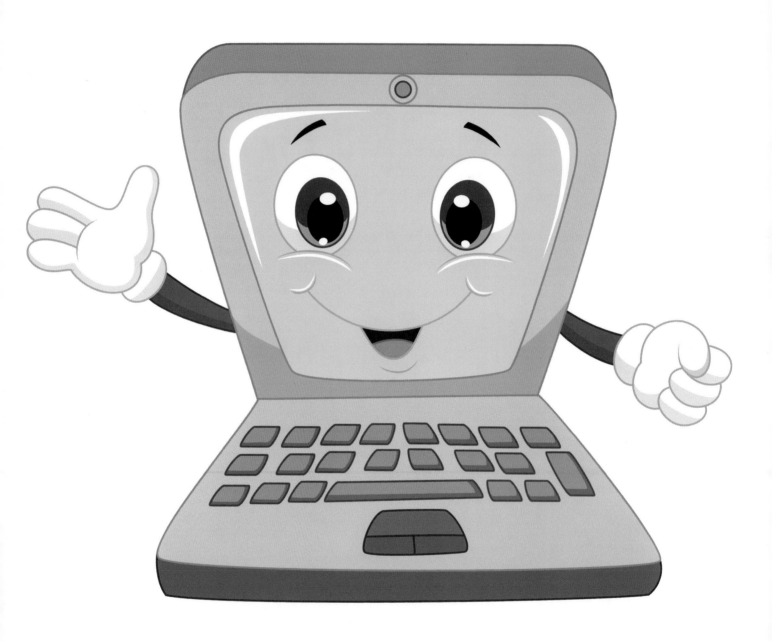

SOLID

Solids Properties of the States of Matter and the Kinetic Theory of Matter Shape Volume Particles are arranged in a fixed pattern hence they have a definite shape.

We know when matter is a solid when they keep their shape unless they are broken. Do not flow or pour liquids.

LIQUID

Liquids Shape Volume Particles are arranged in the form of clusters, with little intermolecular spaces between them. They do not have a definite shape since they take the shape of the container. Because of little spaces between the particles, the volume of liquids is definite.

We can tell when matter is a liquid when they do not keep their own shape; they take the shape of the container they are in.

GAS

GASES Shape Volume Particles are spaced wide apart, they occupy the shape of the container. Since they move randomly and occupy the shape of the container, they have no definite volume.

There are other gases besides the gases in air. Most gases including the gases in air are invisible—you simply cannot seem them.

Printed in Great Britain
by Amazon